Josef

CIRCLES

CAMBRIDGE UNIVERSITY PRESS

Published by the Syndics of the Cambridge University Press
Bentley House, 200 Euston Road, London NW1 2DB
American Branch: 32 East 57th Street, New York, N.Y. 10022

© Cambridge University Press 1967

Library of Congress Catalogue Card Number: 66-16668

ISBN: 0 521 05744 2

First published 1967
Reprinted 1973

First printed in Great Britain by Jarrold & Sons Ltd, Norwich
Reprinted in Malta by St Paul's Press Ltd

Circles and Straight Lines

Can you draw a circle with a ruler?

'Of course not,' you say to yourself, 'a circle is round and a ruler is straight.'

Try this—mark a point on a sheet of plain paper. Take a ruler and put one edge so that it goes through the point, then draw a line along the other edge.

Fig. 1

Repeat this twice more using the same point. Perhaps you have something like this:

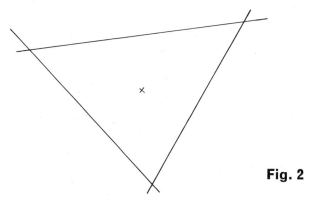

Fig. 2

Not much like a circle is it? It is only a triangle so far. Put in some more straight lines in the same way (Fig. 3).

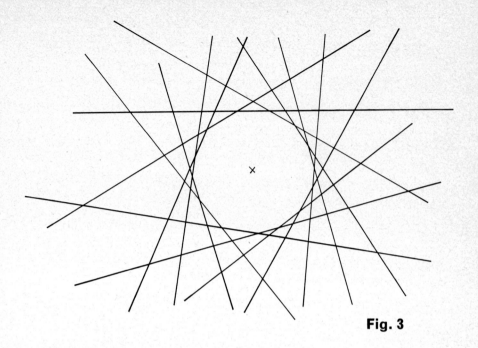

Fig. 3

Do you see how it is beginning to change shape? Go on putting in straight lines until you cannot possibly find space for any more. What shape do you see in the end?

This is something of an illusion or trick in vision. If it were really possible for you to put in an unlimited number of these lines then you would have a circle. As it is you have drawn so many that your eye cannot distinguish them and seems to see them forming a circle.

All the lines you have drawn are called TANGENTS to the circle they form. Notice that they all touch the edge of it but none of them actually goes inside it. How many tangents are there to one circle?

Now draw another circle, in the same way, quite close to your first but using a ruler of a different width. Use a coloured pencil to go over any lines which are tangents to *both* circles. How many are there? You will probably find that they look something like those shown in red in Fig. 4.

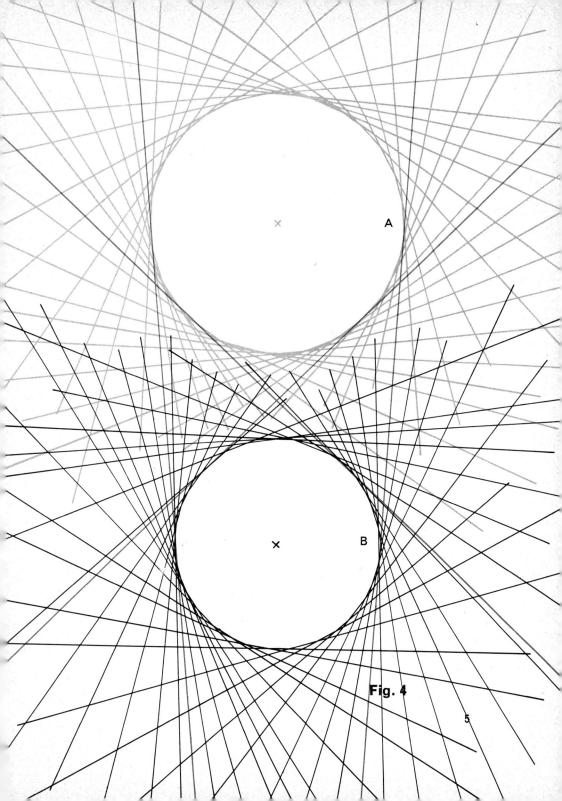

Fig. 4

The set of tangents which are tangents to *both* circles is called the INTERSECTION of the two sets of tangents. There is a special symbol for the intersection of two sets. It is written ∩. If we write T_A for the set of tangents to circle A and T_B for the set of tangents to circle B then we can write $T_A \cap T_B$ (read 'T_A intersection T_B') to indicate the intersection. In Fig. 4 the set $T_A \cap T_B$ has 4 members. Fig. 5 shows just circle A and circle B with the 4 members of the set $T_A \cap T_B$. Each of the tangents is known as a COMMON TANGENT to the two circles.

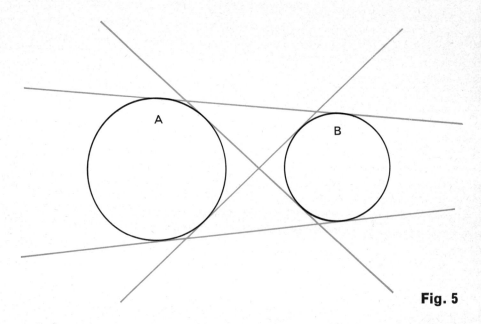

Fig. 5

Supposing the two circles cut into each other: how many members of $T_A \cap T_B$ would there be then? Draw a diagram.

Use a penny and a smaller coin to show two circles related in other ways than apart, as in Fig. 5, or cutting each other. Altogether there are five basically different ways of arranging two circles on a flat sheet of paper. Draw diagrams to show the members of $T_A \cap T_B$ for each way. Does it make any difference if the circles are the same size? Investigate this as fully as you can with drawings and explanations.

Now draw one straight line on your paper and then draw as many circles as possible which have this line for a tangent. Suppose we label the line '1' and write $C_1 =$ the set of all possible circles touching line 1. How many members are there in C_1? Is there any pattern about them? Draw a second straight line parallel to the first. We shall label this line '2' and write $C_2 =$ the set of circles touching line 2. This can be shortened further to $C_2 = \{$circles touching line 2$\}$. Are there any circles which belong to $C_1 \cap C_2$? In other words, do any circles have both line 1 and line 2 as tangents? Fig. 6 shows one possibility.

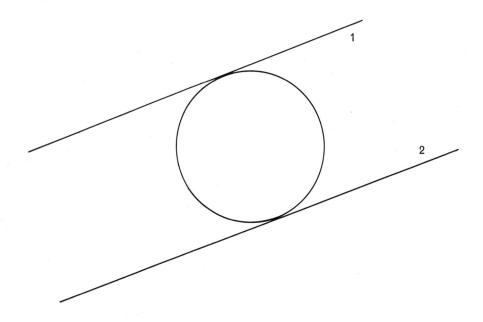

Fig. 6

Draw as many members of $C_1 \cap C_2$ as you have room for. **How many would there be if it were possible to put them all in?** Write down anything that you notice about them. Mark in all the centres and join them up. They form another line which we call the **LINE OF CENTRES**. What relationship does it have with line 1 and line 2?

In Fig. 7 we have two lines marked 3 and 4 which cut each other, and there are also some circles labelled a, b, c, d, e, f.

In the same way as before we have:

$$C_3 = \{\text{circles touching line 3}\}$$
$$C_4 = \{\text{circles touching line 4}\}$$

How would you complete $C_3 \cap C_4 = \{\ldots\ldots\ldots\ldots\ldots\}$?

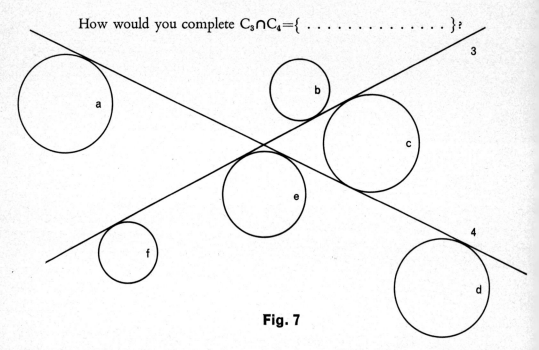

Fig. 7

All the circles in Fig. 7 belong to at least one of the sets C_3, C_4 and $C_3 \cap C_4$. Make a list showing which circles belong to which sets, e.g. circle 'a' belongs to C_4.

Now draw a large diagram putting in as many circles as you can which belong to $C_3 \cap C_4$. You should find that they form themselves into four distinct sections or subsets. Mark in and join up the centres of the circles. How many lines of centres are there? Write down anything that you notice about these lines.

In Fig. 8 you will see that another line, 5, has been added to the diagram. Draw a similarly placed line on your last drawing. Can you find any circles belonging to the set C_5={circles touching line 5}? You should be able to find four circles in $C_3 \cap C_4$ which are also in C_5. Go over them all in a bright colour and see what you can find out about their centres.

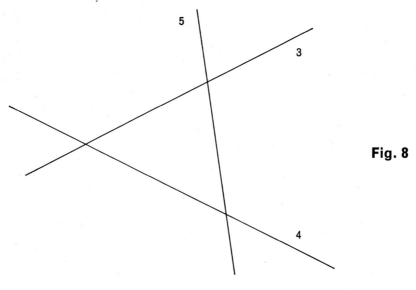

Fig. 8

Circles and Points

Mark a point on your paper with a small cross. Now draw as many circles as you can which go through this point. If we label this point P then all the circles can be described as belonging to set C_P where C_P={circles passing through point P}. How many members are there in C_P altogether? Is there any special pattern about them?

Now make a second point Q on your paper and draw in many members of C_Q={circles passing through point Q}. Are there any members in $C_P \cap C_Q$?

9

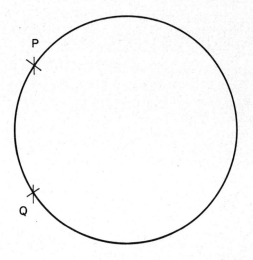

Fig. 9

Fig. 9 shows one of the members of the intersection. Use a coloured pencil to draw in as many members of $C_P \cap C_Q$ as possible. Mark all their centres and join them up. What do you notice about the line of centres? Suppose that you mark one more point R so that P, Q and R make a triangle. How many of the circles in $C_P \cap C_Q$ will go through point R?

Some of the most interesting sets of circles are the ones which are the intersection of two sets, that is, they have two rules to obey. For example in Fig. 10 you see some of the circles from the sets we have already studied.

Fig. 10

Although each of the sets in this figure has really got an unlimited number of members nevertheless they do form some sort of pattern mainly connected with their line of centres.

Fig. 11 shows a set of circles (in black) and one circle on its own (in red) which is not a member of the same set.

What are the rules of membership of the black set?

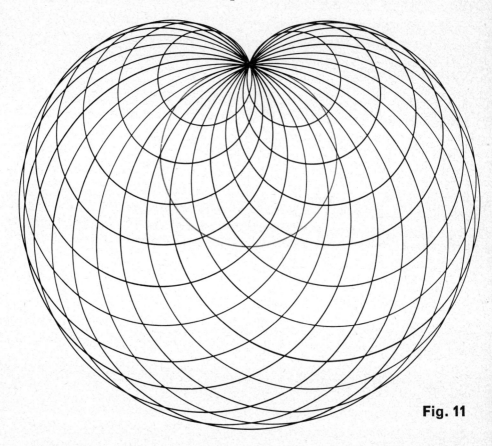

Fig. 11

If you have any difficulty in deciding then look at the next diagram which shows the red circle, which is a sort of controlling circle, and just one of the black ones.

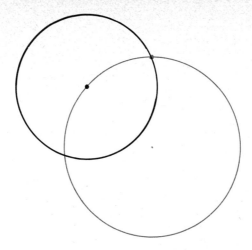

Fig. 12

Try to build up this set of circles in a diagram of your own. When it is finished we see another optical illusion. A shape rather like a circle with a dimple (called a CUSP) has been made by the black circles. This shape is known as a CARDIOID. Find a good dictionary in the library and look up other words which begin with 'cardi-'. How do you think this shape got its name?

The shape formed by the circles in the next diagram, Fig. 13, is rather like a cardioid. Study it carefully and try to decide how this set of circles is like those forming the cardioid and how it is different.

The curve formed round the outside of this set of circles is called a LIMAÇON—this comes from a French word meaning a snail shell. Try to draw this set of circles for yourself.

On the cover of this book you will see another set of circles which form a limaçon. Can you explain why it is that the base circle is completely inside the loop of the limaçon?

So far we have only studied this family of curves when the fixed point has been on the edge of or outside the base circle. Draw sets of circles where the fixed point is inside the base circle and see what happens to the cusp. In particular include one diagram where the fixed point is in fact the centre point of the base circle. What happens to the set of circles in that case?

13

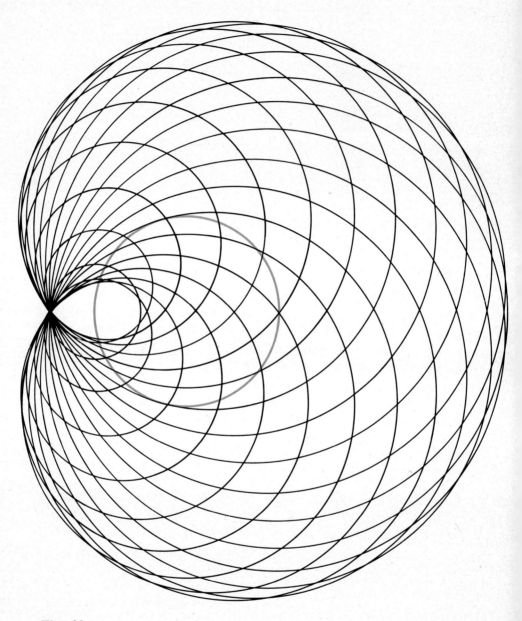

Fig. 13

Have you ever noticed the shape that is formed by the reflection of light on to the tea in your teacup? That curve is part of a near relation of the cardioid and is called a NEPHROID. The shape has two cusps and can be made by drawing the set of circles shown in Fig. 14.

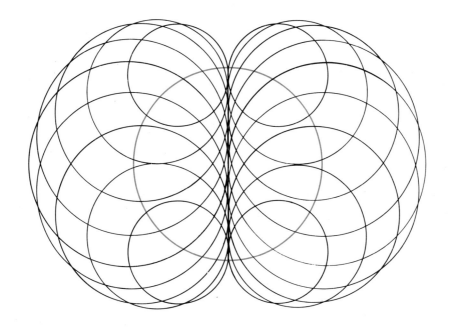

Fig. 14

The prefix 'nephr-' is used in connection with the kidney and as the curve is kidney-shaped you can see how it gets its name. In the diagram there is a base circle, in blue, and all the black circles have their centres on it. A line through the centre of the base circle is also shown. What is the relationship between this line and the circles forming the nephroid? Once you have discovered this you will be able to draw this set of circles for yourself.

You can invent many variations on these sets by splitting the base circle up into a number of equal parts. For example the trefoil shape in Fig. 15 is made by taking three equally spaced lines from the centre of the base circle and then making all the other circles touch the nearest line.

Make up some more for yourself and prepare a wall display.

Fig. 15

All the curves that you have been making are formed by certain relationships between other shapes. In this chapter all the curves were round the outside of a set of circles. The book began by making a circle inside a set of lines. In either case one shape enclosed, or was enclosed by, other shapes. When a curve is made in this way we call it an ENVELOPE. Thus our cardioids, limaçons and nephroids were envelopes of circles and the circle in the first part was an envelope of tangents. The word 'envelope' means to wrap, hide or conceal something, so you can see why it is used to describe the curves you have drawn and how an envelope for letters gets its name.

Folding Circles

Have you ever used filter papers in your science lesson? If so you will know that they are circular pieces of paper rather like blotting paper. See if you can get some to use for the work in this chapter. If you cannot then cut out some paper circles instead.

Fold one of your circles in half. Now open it out and fold it in half again making a different crease. Do this many times and write down what you notice about the creases. **They are all called DIAMETERS. The point where they meet is called the CENTRE of the circle.** How many diameters does one circle have? How many of them do you need to find the centre of the circle? Whereabouts on any diameter is the centre of the circle? Check your answers to the last two questions by folding one circle into quarters (Fig. 16).

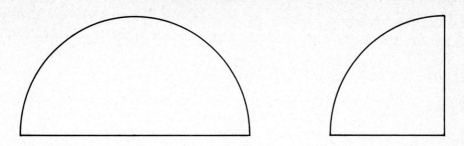

Fig. 16

You probably know that any line from the centre of a circle to its edge is called a RADIUS. What is the relationship between the radius and diameter of a circle?

Take a new circle and find its centre. Now make a fold so that the edge goes through the centre (Fig. 17).

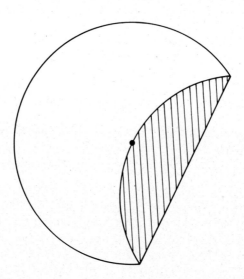

Fig. 17

Unfold and repeat many times so that different points on the edge go to the centre. **What do you notice about the creases this time?** Write down something about their lengths and the shape they form. Straight lines which go across a circle like this are known as CHORDS of the circle. A diameter is a special chord, one which passes through the centre.

If you mark any two points on the edge of a circle how many chords join them together? Suppose you start with three points, how many chords then? Take a circle and mark any four points on the edge; how many chords are needed now to join each point to all the others? Experiment with five, six, seven, eight points and draw up a table like this:

TABLE I

No. of points	No. of chords
2	
3	
4	
....	

Do you notice any pattern about the numbers in Table 1? Use this to see if you can tell how many chords there are in Fig. 18 without counting them.

Fig. 18

These questions might help you if you have not discovered a way yet.

How many points are there round the circle?

How many chords come from each point?

How many chords should this give altogether?

Can you explain why there are only half that number of chords in the diagram?

Now suppose there were 100 points. How many chords would be drawn then?

It is one of the most important jobs of a mathematician to be able to predict what will happen in a certain problem without actually trying it out in practice. This is often done by finding a general formula (a sort of recipe) for all problems of the same type and then applying it to a particular problem. If you were successful in deciding that there should be 4,950 chords if 100 points on the edge of a circle are joined to each other then this is because you had discovered a formula for solving all the problems like this. Discuss your solution with your teacher.

If you want to make an attractive wall display for the last piece of your work try using black paper on a notice-board with drawing-pins for the points and white cotton for the chords. You might like to investigate when you only need one length of cotton to complete the figure and when you need more than one. If you have a lot of points, evenly spaced, then the chords seem to make a series of circles rippling out from the centre. Can you explain why this happens?

By taking particular chords of circles we can obtain a set of tangents enveloping some of the curves we have already met. For example the curve in Fig. 19 was formed by marking 72 evenly spaced points on a circle and then numbering them in a clockwise direction. Each number was then joined to the number which is twice as big. After joining 36 to 72 it was necessary to join 37 to 2 as 2 could be thought of as the 74th point and thus be 2×37. The diagram was complete when each of the 72 points had been joined to its partner. Do you recognize the curve which has appeared?

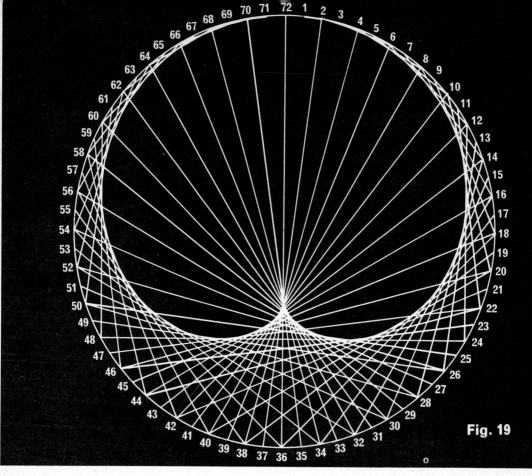

Fig. 19

Experiment on this for yourself taking 36 or 72 points. Try joining any number to three times itself, then four times, then five times and so on. See if you can discover the effect on the curve of changing the multiplying number.

This family of curves is known as the EPI-CYCLOIDS. You have drawn them in two ways so far. Firstly as curves enveloping sets of circles and secondly, in this chapter, as the envelopes of straight lines. They can also be formed in other ways. Fig. 20 shows two circles which are the same size. Imagine that the blue one is kept still and the black one revolves round it. What happens to the point on the black circle which is marked with a cross?

You can set up a rather rough-and-ready model of this if you take two identical cylindrical tins and hold them together by a thin elastic band looped into a figure 8 (see Fig. 21). Fix a pencil to the side of one so that it leaves a line on the paper when that tin is revolved round the other.

What shape is the curve traced by the pencil in one whole turn? Suppose the diameter of the revolving tin was only half that of the stationary tin. What curve do you trace out then? Experiment for yourself with cylinders of different sizes and compare the curves you obtain with those you have found before.

You can read more about these curves and others in a book called *A Book of Curves* by E. H. Lockwood. See if you can get a copy from the library.

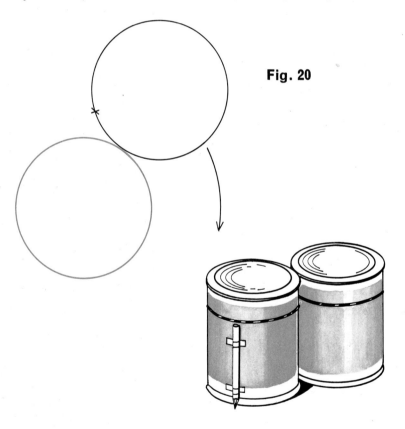

Fig. 20

Fig. 21

Packing Circles

See if you can find 36 coins or counters or discs all the same size. Imagine that you have to pack them all into a flat box of any shape. See if you can work out and draw several ways of packing them. When you have found at least four different ways look at them carefully and see if you can pick out two very distinct ways of setting the circles out in the box. Fig. 22 shows six circles packed into a rectangle and six into a triangle. Apart from the outside shape what is the difference between the two packing methods used?

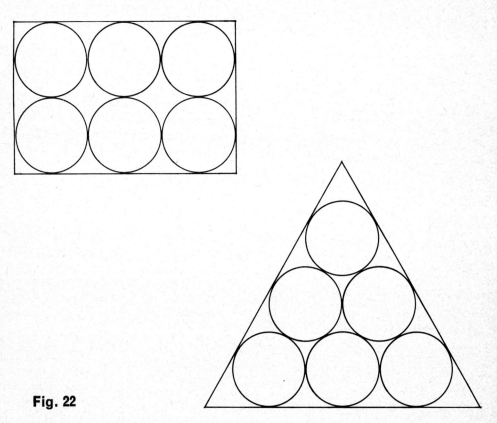

Fig. 22

Imagine that you had to pack some fragile glasses into a crate and you had stiff card to put between them. Mark in on all your drawings of the 36 circles where the stiff card would go. Fig. 23 shows one type of circle pack with red lines where the separators would go.

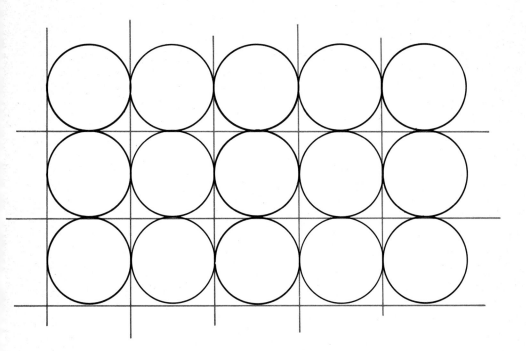

Fig. 23

This way of packing is called SQUARE PACKING and you will quickly see how it got its name.

Fig. 24 shows the other basic type of circle pack. This time the separators make six-sided cells called hexagons. This way of packing is therefore called HEXAGONAL PACKING.

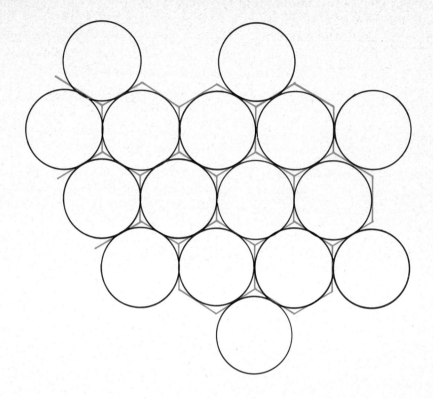

Fig. 24

Which type of packing wastes less space? Why do you think that square packing is very often used when separators are needed for fragile goods and hexagonal packing is used when separators are not needed? Keep your eyes open for examples of circle packing. In particular look for the ones in the following list, make drawings and give explanations of the packing method used.

(a) Bottles in a crate
(b) A box of school chalk
(c) Cigarettes in packets and tins
(d) School milk laid out on a counter or tables
(e) Lead shot or ball-bearings in a tin

(f) Boxes of eggs
(g) Boxes of round fruit
(h) Long pipes stacked on trucks or trailers
(i) Cutting mince pies out of pastry.

In Fig. 25 you see a diagram of a triangular tray which some pharmacists use for counting tablets and pills.

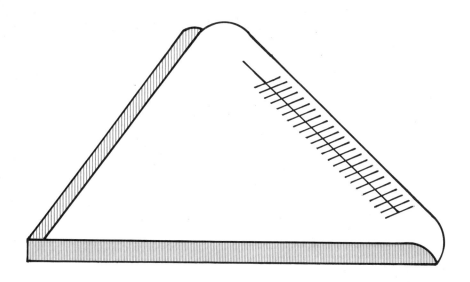

Fig. 25. A pharmacist's tray for counting pills.

How can this be used for counting? Well it depends on the number of rows the pills make when they are shaken down into the corner. Imagine there were only 2 full rows, they would look like this ⊛ ; 3 full rows would be like this ⊛ Describe what happens as the number of rows increases and fill in a table like Table 2 on the next page.

27

TABLE 2

(1) No. of complete rows	(2) No. of pills in longest row	(3) No. of pills altogether
1		
2		
3		
....		

The numbers in column (3) are known as the TRIANGULAR NUMBERS for obvious reasons. Where have you seen them before? Turn back to Table 1 on page 19. There they are again.

Do you remember how you made a formula to find the number of chords in that section? By using the same basic formula we can tell how many pills are in the complete rows and then, by adding on any pills that make part of a row, find the total number. Your chemist does not even have to use a formula, as someone else has used it for him and a table of figures, the same as your column (1) and column (3), has been engraved on the open edge of the tray.

Do you see how a knowledge of such a simple pattern of numbers has been used to save a lot of time for an ordinary person in an ordinary job? You might come across something like this when you are at work.

Measuring Circles

Draw some rectangles with one side 2″, some squares with one side 2″, some triangles with one side 2″ and some circles of radius 2″. Which shapes have turned out the same each time?

To draw any circle it is only necessary to know one basic measurement, usually the radius or diameter. Another length you could measure is the curved edge or CIRCUMFERENCE and you will find, when you do some measuring yourself, that this depends on the diameter of the circle. In mathematics we say that there is a one-to-one correspondence between the diameter and circumference of a circle: if we know one of them then we can calculate the other.

Look for about a dozen circular objects. Such things as plates, tins, clocks, circular mats, bowls, buckets, waste-paper baskets, saucepans and lampshades are ideal. Measure their diameter and circumference as well as you can and make a table of measurements like this:

TABLE 3

Object	Diameter	Circumference
Plate	9"	$28\frac{1}{2}''$

As you are filling in the table see if you notice any connection between the diameter and circumference of each circle. If you think you have seen it try to work out roughly what the circumference will be before you actually measure it.

Fig. 26 shows how to set out a graph of these results. The cross marks the position of the measurements of the plate in Table 3.

Draw the graph with all your measurements on it, on squared paper. Have you got almost a straight line of points? Why isn't it exactly a straight line? Check any measurements which seem out of line. Draw in the straight line which goes best through your points. Should the point (0, 0), the origin, be on the line? Why? This graph could be used to estimate the circumference of a circle if you knew its diameter, and vice versa. How?

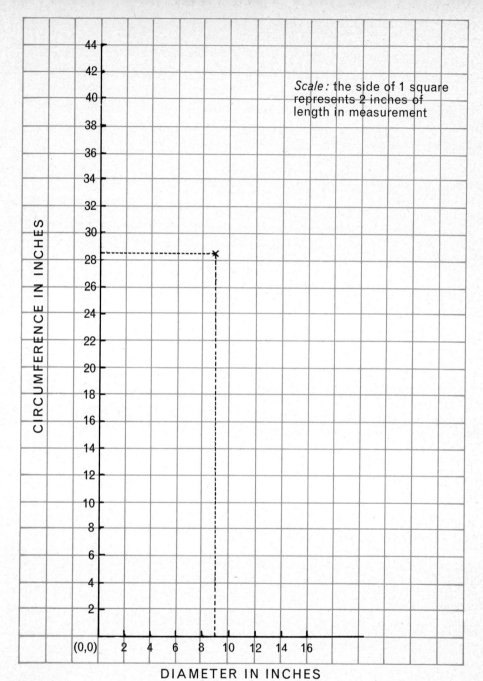

Fig. 26

By now you will have noticed that the circumference is always slightly more than three times bigger than the diameter of the circle. Writing down the exact number of times the diameter fits into the circumference has always presented a problem because it belongs to a class of numbers which cannot be represented exactly by either our fractional or decimal ways of showing parts of whole numbers. In other words there is no fraction which is just right, they are either too big or too small, and if we try decimals then the figures go on for ever and don't even have a repeating pattern. For this reason we give it a symbol instead of trying fruitlessly to write it down fully. This symbol is the Greek letter 'π' (pronounced 'pie').

Despite the impossibility of writing π as a number many people have to use it in their work; architects, scientists and engineers in particular. They get round the difficulty by using an approximation to the right number. Just how good an approximation this has to be depends on the job. For example, if you were buying wrapping paper for a circular object then $3\frac{1}{4}$ times the diameter would be good enough and give you a little to spare.

The Hebrews, who did a good deal of building, used a very rough approximation, an underestimate. Find out what number they used by reading 1 Kings, 7, verse 23, and 2 Chronicles, 4, verse 2, from the Old Testament.

You may find that people tell you that π is $\frac{22}{7}$ (or $3\frac{1}{7}$). The use of this approximation dates back over 2000 years to the Ancient Greek mathematicians. Archimedes (287 to 212 B.C.) calculated that the value of π lay between $3\frac{1}{7}$ and $3\frac{10}{71}$. Look up something about his work and times and see how he did this and find out what other things he studied.

Other approximate fractions have also been used throughout history, notably $\frac{256}{81}$ by the Ancient Egyptians, $3\frac{1}{8}$ by the Romans, $\frac{49}{16}$, $\frac{3927}{1250}$, $\frac{754}{240}$ in India and the East and $\frac{355}{113}$ in China. On 1 January 1964 the following advertisement was put in the papers by the makers of Pye television sets to advertise their new 625-line sets which were to come into use in 1964; it said '$\pi = \frac{1964}{625}$'. What a lucky coincidence for them to find this approximation.

Many people today use 3·142 or 3·1416 for π. It is, however, known to many more decimal places than this, although one cannot imagine that any of them are of real practical use. At the end of the sixteenth century it had been calculated correctly to 15 places of decimals, and this number increased rapidly as mathematicians found quicker ways of working it out. According to present information the lengthiest calculation is to over 100,000 decimal places. This was done using an IBM computer in 8 hours and 43 minutes on 29 July 1961.

Most books on the history of mathematics and many encyclopaedias will tell you more about the history of this number π. Read some of them. Perhaps you could make an illustrated booklet called 'The History of π' for yourself.

ACKNOWLEDGEMENTS

We should like to thank Messrs Pye Ltd for allowing us to quote the advertisement on p. 31; the Wellcome Foundation Ltd for information about the triangular pharmacist's tray; and I.B.M. United Kingdom Ltd for details of the calculation of π to 100,000 places of decimals.

TOPICS FROM MATHEMATICS

Titles in this series

D. S. FIELKER

Computers
Statistics
Cubes
Towards Probability

JOSEPHINE MOLD

Circles
Solid Models
Tessellations
Triangles

TOPICS FROM MATHEMATICS
SOLID MODELS
Josephine Mold

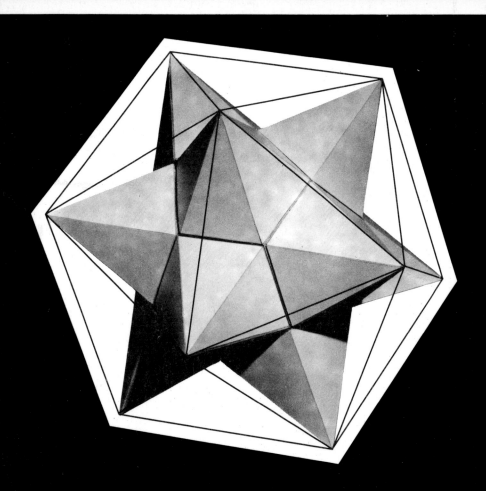

Cambridge University Press